Investigations

Rolling

Patricia Whitehouse

www.raintreepublishers.co.uk
Visit our website to find out more information about **Raintree** books.

To order:
☎ Phone 44 (0) 1865 888112
▤ Send a fax to 44 (0) 1865 314091
▢ Visit the Raintree Bookshop at **www.raintreepublishers.co.uk** to browse our catalogue and order online.

First published in Great Britain by Raintree,
Halley Court, Jordan Hill, Oxford OX2 8EJ,
part of Harcourt Education.
Raintree is a registered trademark of Harcourt
Education Ltd.

Editorial: Nick Hunter and Diyan Leake
Design: Michelle Lisseter
Picture Research: Beth Chisholm
Production: Lorraine Hicks

Originated by Dot Gradations
Printed and bound in China by South China
Printing Company

ISBN 1 844 21553 9
07 06 05 04 03
10 9 8 7 6 5 4 3 2 1

British Library Cataloguing in Publication Data
Whitehouse, Patricia
Rolling
531.1'13
A full catalogue record for this book is available
from the British Library.

Acknowledgements
The publishers would like to thank the
following for permission to reproduce
photographs: Heinemann Library/Que-Net

Cover photograph reproduced with permission
of David Katzenstein/Corbis.

Every effort has been made to contact copyright
holders of any material reproduced in this book.
Any omissions will be rectified in subsequent
printings if notice is given to the publishers.

Some words are shown in bold, **like this**. They are explained in the glossary on page 23.

Contents

What is rolling?

Rolling is one way things can move.

Rolling things turn over and over.

Some shapes roll.

Other shapes do not roll.

Do round things roll?

This ball is on a **smooth** floor.

Gently push the ball.

The ball rolls.

Round things roll easily on smooth floors.

Now the ball is on a carpet.

Gently push the ball.

The ball rolls.

You have to push the ball harder on the **rough** carpet.

Do tubes roll?

Lay a paper tube on its side.

Push the tube.

The tube rolls across the floor.

The side of the tube is **round**.

Stand the tube on one end and push.

The ends of the tube are not **round**.

The tube falls over.

Do pencils roll?

Put a pencil on the floor.

Flick it with your finger.

The pencil rolls across the floor.

But look closely — it is not **round**.

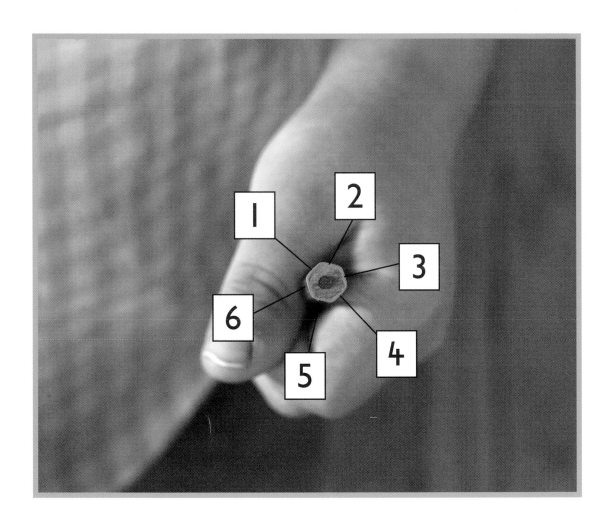

This pencil has six sides.

The six sides give the pencil
a rounded shape.

Things with a rounded shape do roll.

But they need a bigger push.

Do boxes roll?

Put a box on the floor.

Push the box.

The box does not roll.

It does not have a **round** shape.

Now push the box harder.

What happens?

The box **tumbles** over.

It does not roll.

Quiz

Which things roll?

Look for the answer on page 24.

Glossary

flick
to hit quickly

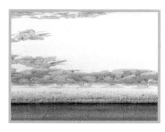

rough
uneven or bumpy surface

round
shaped like a circle or a ball

smooth
even surface that is not bumpy
or rough

tumble
fall over

Index

Answer to quiz on page 22

The ball, paper tube,
and pencil roll.

 CAUTION: Children should not attempt any experiment without an adult's
help and permission.

Titles in the Investigations series include:

Hardback   1 844 21550 4

Hardback   1 844 21551 2

Hardback   1 844 21552 0

Hardback   1 844 21553 9

Hardback   1 844 21554 7

Find out about the other titles in this series on our website www.raintreepublishers.co.uk